RAILWAYS OF LEICESTERSHIRE IN THE TWENTY-FIRST CENTURY

JOHN JACKSON

AMBERLEY

First published 2022

Amberley Publishing
The Hill, Stroud
Gloucestershire, GL5 4EP

www.amberley-books.com

Copyright © John Law, 2022

The right of John Law to be identified as
the Author of this work has been asserted in
accordance with the Copyrights, Designs and
Patents Act 1988.

ISBN 978 1 3981 0269 9 (print)
ISBN 978 1 3981 0270 5 (ebook)

All rights reserved. No part of this book may be
reprinted or reproduced or utilised in any form
or by any electronic, mechanical or other means,
now known or hereafter invented, including
photocopying and recording, or in any information
storage or retrieval system, without the permission
in writing from the Publishers.

British Library Cataloguing in Publication Data.
A catalogue record for this book is available from
the British Library.

Origination by Amberley Publishing.
Printed in the UK.

Contents

Introduction	4
Map Showing the Railway Lines of Leicestershire in the Twenty-first Century	
Leicester	8
Market Harborough	51
South Wigston, Narborough and Hinckley	57
Knighton Junction, Swannington and Leicester Junction (KSL)	61
Loughborough, including Brush Traction	68
Bottesford (Nottingham to Grantham line)	74
Barrow upon Soar, Sileby and Syston	75
Mountsorrel	81
Humberstone Road Sidings, Leicester	84
Melton Mowbray	85
Oakham and Rutland	92

Introduction

Taking a quick glance at a map of England, the county of Leicestershire lies close to its centre. In fact, this landlocked county proudly claims to be 'The Heart of Rural England' on county signs at its road borders. This bold claim may or may not be justified. Earlier this century, however, the Ordnance Survey defined the centre of England as being on the outskirts of Fenny Drayton. This village lies in the south-west corner of Leicestershire, within the district of Hinckley.

A population of just over 1 million people have made the county of Leicestershire their home. Of these, around a third live in the immediate area of the city of Leicester itself.

The city of Leicester enjoys excellent north–south road and rail transport links, with both the M1 and the Midland Main Line running through the area. The Midland Main Line is the route linking London's St Pancras station with the cities of Derby, Nottingham and Sheffield.

Additionally, the city is also served by the east–west route from Peterborough to Birmingham. As far as the rest of the county is concerned, almost all of the passenger stations that survive lie along these two county arteries. The only exception is the village of Bottesford. The station, on the Nottingham to Grantham line, nestles in the far north-eastern corner of the county. The other local stations on this line fall outside the Leicestershire county boundary.

In common with the rest of the United Kingdom, Leicestershire had suffered significant rail closures in the 1960s. These closures came about following the recommendations contained in Dr Richard Beeching's report during that decade. It was published by the British Railways Board and, generally, accepted by the government of the day. Its implementation led to widespread passenger stations and lines being removed from the railway map.

Notably, the closure of much of the Great Central line deprived the Leicester area with its alternative route to London, which ran into Marylebone station. In addition, to the north-west of the county, towns such as Coalville and Ashby-de-la-Zouch were deprived of all rail services with the closure of the line between Leicester and Burton on Trent. The town of Coalville's rail isolation, for example, penalises a population in excess of 30,000 with no immediate access to Britain's rail network.

Three stations between Leicester and Loughborough, at Syston, Sileby and Barrow upon Soar, which closed in 1968, were reopened in 1994. These three reopenings meant that eleven passenger stations in Leicestershire survived into the twenty-first century. By contrast, the list of disused stations in the county totals almost eighty. Several have, though, found new life on preserved lines.

The town of Loughborough is the county's second-largest settlement, with a population of around 60,000. It is the home of Brush Traction's workshops and therefore has a long association with railways and locomotive manufacture in particular.

Two lines worthy of mention, but excluded from the scope of this book, are the Great Central Railway (GCR) and the Battlefield Line. The GCR occupies part of the former Great Central line mentioned above, proudly claiming to be the only double-track, main line heritage railway. The Battlefield Line is operated by the Shackerstone Railway Society and occupies part of the trackbed of the former Ashby and Nuneaton Joint Railway. This Leicestershire line links Shackerstone with Market Bosworth and Shenton.

The adjacent small county of Rutland formed part of Leicestershire for over twenty years towards the end of the last century. Although that is no longer the case, a brief mention of that county's only surviving station at Oakham, and its railway surrounds, is included at the end of this book.

Turning to the freight traffic, the county's industrial past included the mining of coal, chiefly in the north-west of the county and centred on the area around Ashby-de-la-Zouch. When the coal industry was nationalised in 1947, the county had twenty collieries. Today it has none, with the final closure being at Asfordby, near Melton Mowbray, in 1997. Instead, the county's local rail freight traffic is chiefly derived from the delivery of aggregates from several sizeable quarries within the county, and the important ones at Mountsorrel, Bardon Hill and Cliffe Hill, in particular. The area is rich in granite, with Mountsorrel particularly renowned for its pink rock that is much sought after in the UK construction sector. Mountsorrel quarry's operators, Tarmac, send consignments to a wide variety of destinations across the UK and in particular to the south-east of England where there are no naturally occurring supplies of such stone. In addition, the company claims to supply approximately 70 per cent of all Network Rail's ballast for maintaining the country's 20,000 miles of railway track.

The quarries at Bardon Hill and Cliffe Hill are both situated on short spurs from the line that once carried passengers between Leicester and Burton upon Trent. The line is retained as a freight only one, primarily servicing these two customers in the Coalville area. As I write these notes there continue to be aspirations to see a passenger service restored. Only time will tell.

As well as these important local operations, a variety of longer-distance freight traffic passes through the county using the north–south and east–west arteries mentioned earlier. North–south freight traffic benefits from the choice of two routes through the area, either via Leicester station or the longer and more circuitous route via Oakham and Corby. All services have benefited from the recent reinstatement of a four-track railway, on the Midland Main Line further south.

This improvement work has been completed, along with electrification between Bedford, Kettering and Corby. The proposed full electrification of the MML northwards beyond Kettering to Sheffield has since been shelved. The railways of Leicestershire therefore remain an entirely non-electrified one, at least for the foreseeable future, subject to the caveat below.

Long-distance, east–west freight services can utilise the link between the West Midlands, Nuneaton, Leicester and Peterborough in order to reach East Anglia, and the port of Felixstowe in particular. This gives an attractive alternative to rail freight operators instead of using the busy lines via the London suburbs.

Finally, a stretch of 13 miles of railway is retained near the town of Melton Mowbray, accessed via a spur from this east–west artery. The Old Dalby Test Track, as it is commonly known, runs between Melton Mowbray and Edwalton, utilising part of the former Midland Railway route between Kettering and Nottingham. This line is used for testing a wide variety of new trains and, ironically, incorporates the only stretch of electrified railway in the county not for use by the general public but for these testing purposes. The project is known by Network Rail as the Rail Innovation and Development Centre, Melton, and includes sidings and depot facilities close to the village of Asfordby.

Our journey around the county starts in Leicester itself with a detailed look at the modern railway scene in the station vicinity. This is followed by an overview of the county's railways, starting in the south at Market Harborough and moving clockwise round to Melton Mowbray. Finally, we take a brief look at Oakham and Rutland's surviving railway.

As always, I hope you enjoy your browse through the pages that follow as much as I have enjoyed compiling them.

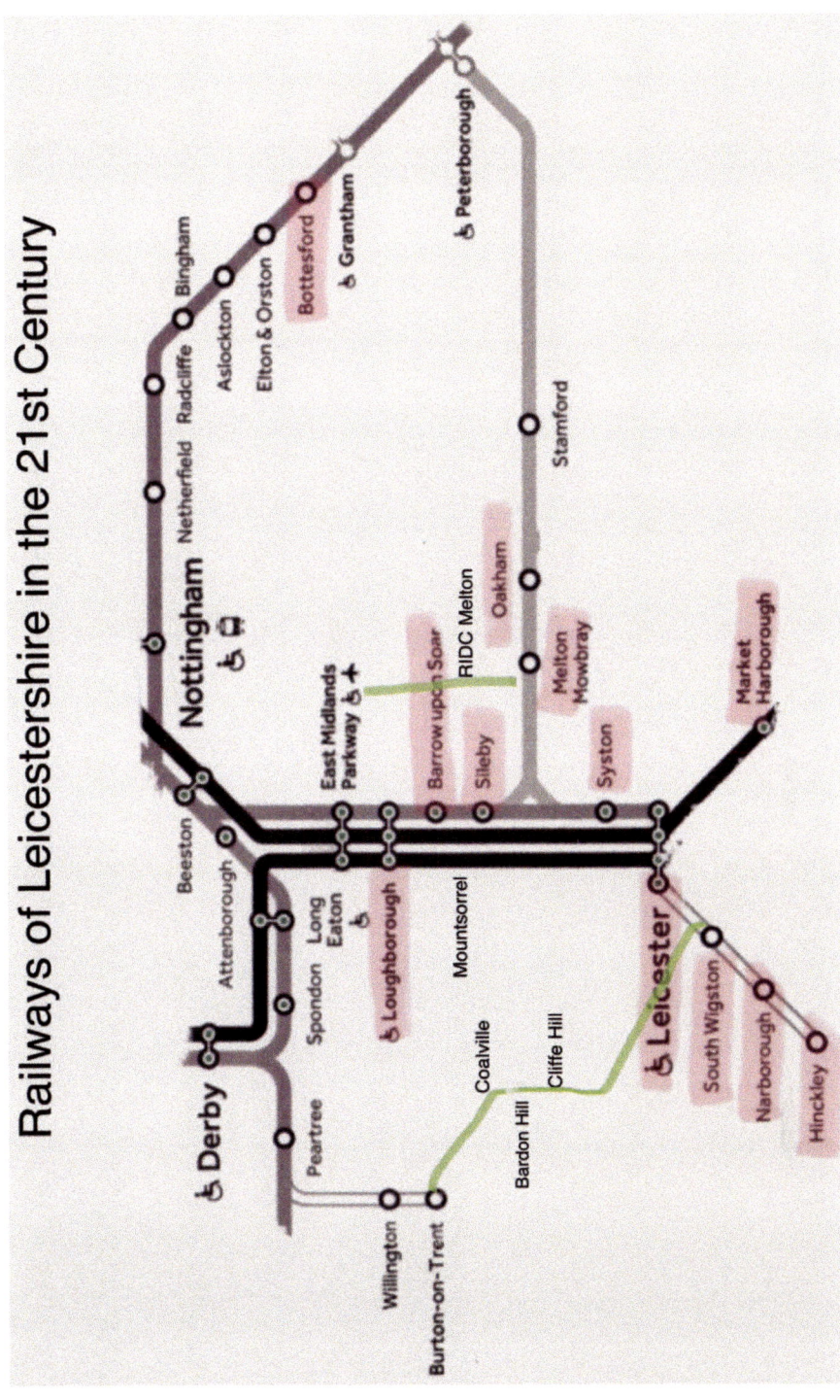

This railway map shows Leicestershire and Rutland's passenger stations shaded in red. These twelve are the only twenty-first-century survivors in a county that once boasted around 100. The green lines indicate two important non-passenger lines, serving the quarries close to Coalville and the test centre (RIDC) close to Melton Mowbray.

Leicester

We commence our journey around the county of Leicestershire's railways in the twenty-first century in the immediate area of Leicester station. This is the 2021 view of the city's station entrance on London Road. The main railway lines tunnel under the road here and the station platforms are below street level.

This view of Leicester's four-platform station is taken from the footbridge at the north end of the station. With platforms 1 and 4 out of picture to the right and left, five-car Meridian No. 222008 is seen leaving platform 3 on a service to London St Pancras.

The Intercity 125 high-speed trains (HST) have been the mainstay on the Midland Main Line for almost forty years. Power car Nos 43047 and 43045 head north from Leicester on 3 March 2020 on a London St Pancras to Nottingham service. This was to be the class's last full year in service with East Midlands Railway.

At the turn of the twenty-first century, the Midland Main Line livery of teal and tangerine was the norm. Power car No. 43048 is seen waiting to head north from Leicester station.

By 2003, the line's HST operators, Midland Main Line, had revealed a new corporate identity. Sporting the company's new blue livery, No. 43104 calls at Leicester on 11 March that year.

During 2003 and 2004, Midland Main Line operated additional passenger services between London St Pancras and Manchester Piccadilly. This was at the request of the Strategic Rail Authority while engineering work was carried out on the West Coast Main Line route from London Euston. Additional HSTs were hired in for this purpose, including First Great Western's power car No. 43005, seen at Leicester on 30 July 2003.

Having taken over the franchise in 2007, East Midlands Trains (EMT) operated services on the Midland Main Line for eleven years. They were superseded by Abellio's East Midlands Railway in 2019. To mark the end of their franchise, a pair of HST power cars, including No. 43081, were branded with 'Eleven Years of East Midlands Trains'.

The second of the branded power cars, No. 43050, is seen on the rear of a Nottingham to London St Pancras service as it departs from Leicester on 16 August 2019.

At the end of 2017, passenger operator Grand Central withdrew its small fleet of HST sets, with the stock transferring to East Midlands Trains. One of the six power cars involved, No. 43484, is seen at Leicester on 10 June 2019. It is the rear power unit on a Nottingham to St Pancras service.

A more unusual move on 30 August that year saw 'back-to-back' power cars, Nos 43480 *West Hampstead PSB* and 43423 *Valenta 1972–2010*, heading through Leicester on a Derby to Cricklewood proving run.

The fleet of high-speed train sets, operated by East Midlands Trains (EMT), has been complemented by a fleet of twenty-three Class 222 Meridian units, first introduced by the company's predecessor, Midland Main Line, in 2004. On 8 June 2016, seven-car example No. 222004 arrives at Leicester on a London St Pancras to Sheffield service.

A further four units were transferred from Hull Trains in 2009/2010, and today's fleet of twenty-seven units is split into four-, five- and seven-car sets. On 29 July 2021, four-car unit No. 222101 approaches Leicester on a London service. The unit is now sporting East Midlands Railways' aubergine livery.

While these Meridian units primarily work London services, they have made appearances on the local services from Leicester to Lincoln via Loughborough and Nottingham. On a miserable January day in 2017, No. 222101 waits to form a service to Lincoln. The length of these units was incompatible with the short platforms at several of the stations served.

Midland Main Line's Meridians were manufactured by Bombardier, responsible for the earlier Voyager trains for both Virgin and Cross Country. These Voyagers are only occasional visitors to Leicestershire. On 17 July 2017, Cross Country example No. 221139 is seen on diversion through Leicester station when its normal route through Burton on Trent was blocked by a train failure.

Leicester station sees a variety of stock from other passenger operators making an appearance. On 2 February 2012, for example, Adelante set No. 180106 heads north through platform 1 on a move from Old Oak Common, West London, to Wabtec's recently acquired workshops at Kilmarnock, Scotland.

Heading in the opposite direction on 3 March 2014 was sister Adelante No. 180112, operated by Grand Central. For some unknown reason, this unit was routed via Leicester on a move from Heaton depot in Newcastle to the depot at Longsight in Manchester.

Several years later, these Adelante units were to become regular passenger performers through the county, with four sets being transferred to East Midlands Railway (EMR) from Hull Trains in 2020. On 14 June 2021, No. 180111 heads north from Leicester on a St Pancras to Nottingham service.

With the last of EMR's HST sets withdrawn from regular service in May 2021, the Meridians and Adelantes now jointly operate the services between London, Nottingham and Sheffield. The new norm is reflected in this view from the north end footbridge on 2 August 2021, with Nos 222023 and 180109 side by side in the centre platforms. Fleet replacement with new Class 810 units is expected in 2023.

EMR's long-distance services to and from London St Pancras are complemented by a local service operating northwards through the county to Loughborough and beyond. On 16 October 2018, the graffiti on the side of two-car diesel unit No. 156401 brings a little unexpected colour to the scene as it arrives in Leicester's platform 4 on a service from Lincoln.

The early 1990s saw a number of improvements to the local service from Leicester northwards to Loughborough, including reopened stations, covered later in this book. An hourly service was introduced under the 'Ivanhoe Line' branding. On 18 January 2017, No. 156498 leaves on a late afternoon service.

A pair of Class 153 single units were often provided rather than the two-car Class 156 units. On 13 December 2016, for example, No. 153374 leads a sister unit on a Lincoln departure.

Space was at a premium on this service, and its return, on 13 March 2017. Single-car unit No. 153313 was provided by EMT for this service between Lincoln and Leicester and back.

In more recent years, Class 158 units have also worked on the Ivanhoe Line services to Loughborough and onwards to Lincoln. On 16 October 2018, No. 158785 waits to leave Leicester.

Sister unit No. 158889 was loaned to the East Midlands Trains' fleet in May 2015, from Stagecoach's South West Trains franchise. Its transfer had been made permanent, and the unit rebranded, by the time this photo was taken on 6 June 2018.

Passenger services on the east–west route across the county are in the hands of Cross Country Trains. An hourly long-distance service between Birmingham New Street and Stansted Airport is provided by their fleet of Class 170 units. On 3 September 2018, No. 170397 waits in platform 2 on a service bound for Stansted. These services are complemented by an additional hourly service linking Birmingham to Leicester, calling at local Leicester stations.

All these eastbound services call at the Leicestershire town of Melton Mowbray and the Rutland town of Oakham as they make their journey to Stansted Airport via Peterborough and Cambridge. On 6 December 2018, No. 170109 leaves Leicester on the 10.18 service to Stansted.

As we have already seen with the Adelantes earlier, a variety of units from other operators pass through Leicester station from time to time. On 16 December 2017, South Western Railway's No. 158890 is seen on the goods loop adjacent to Leicester station platforms. It is returning from Brush Traction's workshops at Loughborough to its home depot at Salisbury.

Heading in the opposite direction on 31 January 2017, a London Midland diesel multiple unit is seen passing through on a move from its home depot at Tyseley, Birmingham, to the Brush workshops at Loughborough.

The cancellation of the electrification plan for the county's main railway line does not prevent electric multiple units from making an occasional appearance. The Class 345 Crossrail units are seen on moves to and from their London bases. On 30 August 2019, No. 345063 is seen being dragged by a Class 37 diesel loco from Old Oak Common, West London, to Worksop for further storage.

A diesel oddity that appears here from time to time is Network Rail's unit No. 950001. This Track Measurement Unit was built in 1987 specifically for their needs. It uses the same bodyshell as the early Class 150 Sprinter passenger units. It is seen waiting on the goods loop on 12 November 2014.

In addition to the regular passenger services, Leicester station sees a variety of freight movements within the station area. The substantial granite quarry at Mountsorrel lies about 7 miles to the north of the station. On 15 February 2018, No. 66015 heads north with a rake of empty wagons returning to the quarry. This consignment of stone had earlier been unloaded at Elstow, on the outskirts of Bedford.

On 24 February 2020, No. 66031 heads for Mountsorrel with a rake of empty wagons, this time from Crewe Basford Hall Yard. Despite the EWS livery, the loco is in use with Direct Rail Services (DRS). The loco will return to Crewe with a loaded rake of ballast for use by Network Rail.

This working often brings one of DRS's fleet of Class 68 locomotives to the area. On 9 March 2018, it is the turn of No. 68004 *Rapid*, seen heading through platform 1.

In this view on 24 April 2019, DRS have allocated sister loco No. 68031 *Felix* for this working. Despite its dedicated TransPennine Express livery, it is consigned to this Network Rail duty that day, caused by a delay in implementing their passenger duties across the Pennines.

DRS also handle Network Rail's ballast requirements at the substantial yard at Carlisle. The returning working from Mountsorrel to Carlisle is usually routed north from the quarry. On 6 March 2018, however, No. 66304 is seen approaching Leicester with the return loaded working. On this occasion, the train will head north on the West Coast Main Line after heading westwards to Nuneaton.

GB Railfreight also handle stone workings to and from Mountsorrel from various rail yards around the country, including Eastleigh, near Southampton. On 30 December 2017, their Class 66 locomotive No. 66775 heads yet another rake of Network Rail yellow box wagons (IOAs) through the station. The quarry accounts for an estimated 70 per cent of all Network Rail's ballast needs.

The quarry at Bardon Hill is also responsible for a number of stone workings through the Leicester station area. On 24 March 2014, workings to and from that quarry were then in the hands of Freightliner. Their Heavy Haul Class 66 locomotive, No. 66615, is seen heading south towards the station with a return working from Tinsley, near Sheffield.

On 18 April 2016, GB Railfreight's (GBRf) No. 66702 *Blue Lightning* heads through the station on another rake of empties returning to Bardon Hill quarry. This time the working is from Ferme Park, North London. Both these workings will require the locos to run round their trains in the loop at Knighton, just south of the station. They can then access the branch line to Bardon.

The significance of the aggregates traffic on today's railway is further demonstrated by this selection of workings returning to the quarries further north, in Derbyshire. On 2 March 2020, Freightliner's No. 66607 passes through the platform on a working of empty hoppers returning from Tarmac's aggregate handling facility at West Thurrock, Essex. Its destination is the Derbyshire quarry at Tunstead.

The same quarry is the destination for this rake of GBRf wagons, which have been hauled by No. 66722 *Sir Edward Watkin* as they head north for loading on 14 June 2021.

The workings between Dove Holes Quarry at Peak Forest and the Cemex receiving depot at Bletchley are among the longest-standing freight diagrams in the UK. In 2016, these trains were operated by DB Cargo. On 2 September that year, No. 66115 takes the goods loop as it heads north. By then, this working was making use of HTA wagons, rendered redundant by the downturn in coal traffic.

These workings are now in the hands of GBRf, with No. 66761 *Wensleydale Railway Association 25 Years 1990–2015* in charge on 14 December 2018. They, too, have made use of their own redundant coal wagons on these stone workings.

A different type of ballast working makes an occasional appearance through Leicester. Network Rail's High Output Ballast Cleaner is one of the longest trains on the rail network, with a loco at each end. On 19 October 2016, a pair of Freightliner Class 66 locomotives are in charge of this working as it approaches Leicester station with No. 66547 leading.

The rear loco on this working from Stapleford is No. 66621. By the time it reaches the station platform, the lead loco is out of sight. On this occasion the pair and their train are heading for a worksite near Luton.

In addition to the numerous stone workings, the Leicester area sees an increasing amount of container traffic to and from the port of Felixstowe. Despite the option of routing via the London suburbs, both Freightliner and GB Railfreight choose the option to reach Felixstowe by travelling via Peterborough and across East Anglia to Ipswich. On 17 January 2017, No. 66416 waits to head north from the goods loop on a Freightliner working from Lawley Street terminal in Birmingham.

On 18 April 2016, sister loco No. 66556 approaches the station on a working from Felixstowe to Crewe. It will join the West Coast Main Line at Nuneaton then head north to its destination.

This working occasionally produces a Freightliner Class 70 locomotive, rather than a Class 66. This was the case on 16 September 2016 when No. 70011 was in charge.

GB Railfreight operate several daily container trains between Felixstowe and the West Midlands. On 13 March 2017, No. 66779 *Evening Star* passes through Leicester on an intermodal from Hams Hall Rail Freight Terminal, near Coleshill. This complex, operated by Associated British Ports, is the country's busiest inland rail freight terminal.

On 11 July 2014, No. 66710 *Phil Packer BRIT* is routed via the goods loop rather than straight through the platform. It will have to wait for the local passenger service to clear the single-line section north of here before it can continue on its journey north then eastwards to Felixstowe.

On 9 February 2017, it's the turn of DB Cargo's No. 66136 to wait on the goods loop for the signal to proceed. This container train from Burton on Trent has taken a circuitous route via the West Midlands in reaching this point. It will operate north to Syston and then head eastwards to Felixstowe.

Another longstanding freight flow through Leicester is the DB Cargo working of empty steel wagons returning from Corby to Margam, in South Wales. On 10 October 2014, No. 66150 approaches the station on this working.

On 26 May 2016, No. 66077 *Benjamin Gimbert GC* is seen in charge of a longer rake of empty steel wagons on an eight-hour journey from Corby back to South Wales. This steel coil traffic to the town's pipe works is all that remains since the closure of the once huge steelworks back in 1980.

Leicester sees various other freight services, provided by the majority of the main freight operators. For a number of years, Colas Rail handled the aviation fuel flow from Lindsey, Humberside, to Colnbrook, for Heathrow Airport. On 28 March 2018, No. 60056 takes Leicester's goods loop on the return working with the empty tanks.

On the same day, GBRf's large logo blue celebrity No. 66789 *British Rail 1948–1997* also takes the goods loop with a rake of empty tanks returning to Immingham, Humberside, from unloading at Puma Energy at Theale, Berkshire.

Two years later and, on 29 July 2021, the same working is in the hands of DB Cargo. Their Class 60 locomotive, No. 60015, is given the road through platform 1 as it returns its rake to Humberside.

Many of these Class 60 locomotives have been regular visitors to Leicester since being built at nearby Brush Traction at Loughborough thirty years ago. The same loco, then named *Bow Fell* and carrying Transrail livery, passes the same platform. This unidentified tank working dates back to the turn of the century.

Another important Freightliner customer is Hope Cement Works in Derbyshire. Most weekdays a loaded cement working passes Leicester on route to the terminal at Theale in Berkshire. On 18 January 2017, No. 66601 *The Hope Valley* heads the empty tanks through Leicester's platform 2 on their return working to Earles Sidings at Hope, Derbyshire.

Two of Colas Rail's small fleet of Class 66 locomotives were used on 26 January 2016 on a ROBEL track machine move. Loco Nos 66848 and 66850 *David Maidment OBE* double-head the machine north from Leicester on a move from West Ealing to Chaddesden, Derby.

For a number of years, veteran Class 20 locomotives were regularly used on moves involving new units for London Underground (LUL). These workings involved movement between manufacture at Derby, unit testing on the test track at Old Dalby and delivery to West Ruislip, the point of entry for the London Underground. On 2 March 2016, Nos 20314 and 20096 are seen at the far end of an LUL set waiting in Leicester's goods loop.

A total of four Class 20 locomotives, two at each end of the LUL stock, was used in each movement. On 5 July 2013, No. 20314 is seen in use again, on this occasion with No. 20311, on the northern end of this working. Both these locos were on hire from Harry Needle Railroad Company.

Another regular contract involves pairs of locomotives hauling the annual rail head treatment trains (RHTT) during the leaf-fall season. On 19 October 2015, DB Schenker's Nos 66101 and 66197 are at either end of the two water cannons working a circuit based on Peterborough. A second diagram covering the Midland Main Line is centred on the company's depot at Toton, Nottinghamshire.

Leicester station frequently sees one-off wagon moves. A GB Railfreight example of this occurred on 25 April 2019, when No. 66776 *Joanne* was called on to move a short rake of assorted wagons between Wellingborough Yard and March, in Cambridgeshire.

The Midland Main Line is regularly used for the movement of a variety of Network Rail test trains between their Derby base and their work locations. On 4 March 2016, Brush Type 2 locomotive No. 31233 is seen on the rear of a working returning from Rugby to Derby.

On 10 March 2020, it is the turn of Colas Rail's English Electric Type 3 locomotive, No. 37116, to appear on another similar working. On this occasion, the loco is on the rear of a positioning move from Derby to Hither Green, South East London.

Class 67 locomotives have also been regular performers on these duties. On 11 December 2014, No. 67015, then named *David J. Lloyd*, is on the rear of a working from Derby to Worcester Shrub Hill.

Pairs of GB Railfreight's Class 73 electro-diesel locomotives are often used on workings within the former Southern Region of British Rail. On Tuesday 24 May 2016, No. 73962 *Dick Mabbutt* leads one of these workings, with a sister loco out of sight on the rear. The pair will spend the remainder of the week working a number of diagrams centred on Tonbridge Yard, Kent.

Network Rail's Inspection Saloon, No. 975025 *Caroline*, appears regularly while on inspection duties, usually in partnership with a Class 37 locomotive. On 9 March 2018, it is the turn of No. 37407 to be the chosen motive power as it leads the saloon between Derby and Kettering.

Another celebrity machine, this time operating with Freightliner, appeared on 2 March 2020. Their Brush Type 4 Class 47 locomotive No. 47830 *Beeching's Legacy* was seen heading north on a Crewe to Peterborough, and return, route learner. The loco carries the number D1645, the original number carried when first built in 1964.

Light engines movements to and from Brush Traction's works at Loughborough frequently bring visitors through the Leicester station area. On 13 March 2017, GB Railfreight's No. 73961 *Alison* drags Caledonian Sleeper loco No. 92023 on a move from Tonbridge Yard to Loughborough.

Locos carrying the Virgin Trains livery were much rarer visitors to the area. On 8 September 2011, however, three former Virgin-operated Class 57 locos, Nos 57305, 57303 and 57301, pause in the station platform whilst heading to Brush at Loughborough.

The Class 58 locomotives have a long association with the area, right from the time they were built by British Rail Engineering at Doncaster in the early 1980s. On privatisation, the class of fifty locos passed in to the hands of EWS (now DB Cargo). They only just survived in use into the twenty-first century, with most taken out of service in 2002. A pair of locos, Nos 58013 *Doncaster Works* and 58020, pause at the north end of the station on 30 May 2000.

Also in 2000, Brush-built Class 60 locomotive No. 60093 *Jack Stirk* pauses on the goods loop as it awaits the 'dolly' signal. In the privatised hands of EWS, the class was to meet a similar fate to the Class 58s, with around 50 per cent of the 100-strong fleet being out of use by the mid-2000s.

Leicester station has occasionally seen light engine moves involving Network Rail's locomotive pool. On 1 May 2013, veteran machine No. 31105 leads No. 97304 on a move returning the locomotives to Derby.

By the time this photo was taken on 30 August 2002, the General Motors' Class 66 locomotive had replaced most older locomotives within the EWS fleet. Delivery of the entire order of 250 locomotives had been completed by the time No. 66084 is seen here moving on to the depot at Leicester.

The other freight operators have followed the lead of EWS and established Class 66 loco fleets for themselves. On the morning of Monday 8 March 2018, two GB Railfreight examples, Nos 66725 *Sunderland* and 66728 *Institute of Railway Operators*, pass through Leicester station on a light engine move from their depot at Peterborough. They are heading to the freight terminal at Hams Hall to commence their week's duties.

The former depot at Leicester has seen wide-sweeping changes of use over the last twenty years with a number of different rail-related businesses operating from the site, since EWS ceased to use it from around 2007. In 2021, the depot area has seen the return of a Class 08 shunter locomotive. Former Wolverton Works celebrity, No. 08629, is seen shunting barrier vehicles on 29 July that year.

In recent years, many different classes of loco have made appearances at Leicester, some briefly and others languishing for years. In this general view, taken on 4 December 2019, the line-up of heritage locomotives includes Nos 56103, 37608, 37188, 37906 and 57312.

Preserved Class 46 locomotive No. 46045 made a brief stopover at Leicester in September 2016. It is seen here prior to a move to the Midland Railway at Butterley on the 20th. The loco carries its original number, D182, and a 52A Gateshead depot shedplate.

In company with No. 37350, Deltic Class 55 locomotive No. 55002 *The King's Own Yorkshire Light Infantry* arrives at Leicester on 18 July 2016. The pair had earlier left the National Railway Museum at York.

Former West Coast Main Line electric locomotive No. 86702 *Cassiopeia* spent several months at the depot in 2015. It is seen on 4 June that year.

More recently, former East Coast Main Line electric locomotives have been visitors to the depot area. On 10 March 2010, No. 91120 is seen carrying its new Europhoenix livery from passing on the adjacent Midland Main Line.

Many of the surviving Class 56 locomotives have been seen at Leicester during the last few years. On 18 June 2018, this is a typical line-up on the depot with No. 56081 leading Nos 56311, 56031, 56037, 56069 and 56032. In 2021, the shell of No. 56031 was to emerge as GB Railfreight's No. 69001.

English Electric Class 50 locomotive No. 50008 *Thunderer* is a regular at Leicester. The loco is now owned by Hanson & Hall Rail Solutions. This view shows the loco on 29 April 2019 preparing to leave Leicester on a light engine run to Knottingley depot.

Many of today's loco moves to and from Leicester involve movement of passenger operators' rolling stock around the country. This often involves the use of barrier vehicles. On 29 July 2019, a pair of Rail Operations Group's Class 47 locos, Nos 47815 *Lost Boys 68–88* and 47813 *Jack Frost* are about to leave on a barrier vehicle move to Crewe.

The sidings adjacent to platform 4 at Leicester station are frequently used for stabling departmental track machines. For example, on 13 December 2016 these sidings were home to Swietelsky Babcock Rails' No. DR73804 *James Watt*.

Market Harborough

We start our journey around Leicestershire at the county's most southerly railway station, Market Harborough. The town is 16 miles south of Leicester and the first stop on the Midland Main Line (MML) to London St Pancras. The Grade II listed station building dates from 1884 and is seen here on a rare occasion when free of road traffic, Christmas Day 2012.

This photo, taken on 7 June 2018, shows the more familiar view as seen by the passing rail traveller. The Midland Railway building is seen at track level, taken from the London-bound platform.

At that time, the town enjoyed an hourly non-stop passenger service to and from London St Pancras. These Nottingham services were in the hands of East Midlands Trains' fleet of HST sets. Power car No. 43059 is seen here calling on a service towards Nottingham.

Today, services are primarily in the hands of Meridian units. On 16 April 2004, and carrying Midland Main Line's livery of the time, five-car Meridian unit No. 222008 passes Market Harborough on a test run.

This stretch of the MML sees a variety of traction working Network Rail Test Trains from its base at Derby. Many of these are destined for the former Southern Region area of the rail network. For example, on 25 February 2012, a pair of Brush Type 2 locomotives, with No. 31465 leading and No. 31106 on the rear, are seen passing through the station on a working from Derby to Grove Park, in South London.

Heading in the opposite direction on 13 October 2015, GB Railfreight's Class 73 locomotive, No. 73212 *Fiona*, brings up the rear of a working returning to Derby. The train had spent several days working in the area around Tonbridge, Kent.

The Midland Main Line is also used in order for a variety of train operators to move rolling stock to and from railway workshops in the Leicestershire area. On 31 December 2012, Chiltern Railways' unit No. 172103 is seen returning from Brush workshops in Loughborough to its base at Wembley Stadium.

Freightliner's No. 66953 is the chosen motive power for this new unit move on 13 October 2016. It is seen passing Market Harborough dragging electric multiple unit No. 387303 from Derby to Bletchley. The unit will undergo mileage accumulation runs on the West Coast Main Line before delivery to its operators, C2C.

Some of the freight traffic on the southern end of the Midland Main Line (MML) is routed away from Leicester and Market Harborough and operates via Syston, Oakham and Corby. Other traffic uses the MML throughout. This northbound Freightliner working on 13 October 2015, for example, sees No. 66620 returning to Earles Sidings, in Derbyshire's Hope Valley, after unloading at Theale, Berkshire.

In 2019, Network Rail carried out major improvement work in the Market Harborough station area as part of the wider Midland Main Line upgrade. The straighter line through the station platforms is evident in this view looking south as No. 222001 heads north with a St Pancras to Sheffield service.

The station has a new footbridge to improve accessibility from the car parks. On 21 July 2021, No. 222016 calls at the southbound platform on a Nottingham to London St Pancras service.

This is the view looking north on the same day. The notorious curves through the town's platforms have been removed and replaced with straight track. One of East Midlands Railway's small fleet of Class 180 diesel units, No. 180109, calls on a St Pancras to Nottingham service.

South Wigston, Narborough and Hinckley

Three Leicestershire passenger stations survive on the line between Leicester and Nuneaton, Warwickshire. This line, part of the east–west route linking Peterborough and Birmingham New Street, leaves the MML at Wigston Junction, 3 miles south of Leicester. This photo shows the basic street level entrance to the first of these stations at South Wigston. This station was opened in May 1986, around two decades after closure of three stations in the immediate Wigston area. A triangular junction with the MML is retained here to simplify freight movements in the area.

On 21 July 2021, Cross Country Trains' unit No. 170108 passes eastbound on a service from Birmingham New Street to Stansted Airport. The two station platforms are staggered either side of the footbridge in the above photo, and facilities at this unstaffed station limited to a bus-style shelter on each platform.

A further 3 miles west of South Wigston is the station at Narborough. The station was closed by British Railways in 1968, but was restored thanks to local council funding and reopened two years later after local opposition to its closure. This view, taken from Station Road, shows the signal box and level crossing adjacent to the station.

This cross-country route enjoys a number of regular freight workings each weekday, particularly to and from the freight terminals in the West Midlands area. On 29 July 2021, GB Railfreght's No. 66745 *Modern Railways The First 50 Years* heads east with an intermodal from Birch Coppice to the port of Felixstowe.

This Leicestershire stretch of line also sees a regular Colas Rail working, usually hauled by one of their fleet of Class 70 locomotives. The working consists of empty box wagons from Westbury being taken to Midlands Quarry Products' Cliffe Hill Quarry for loading. On this occasion, the loco is No. 70807.

The third Leicestershire station on this line is at Hinckley, around 15 miles from Leicester and close to the county border with Warwickshire. With a population approaching 50,000, it is the third-largest community in the county, behind Leicester and Loughborough. This is the view of the station looking towards Leicester. The public footbridge in the centre is also used to gain access to the platforms.

Another eastbound container train is seen passing Hinckley on 29 July 2021. This time the train has originated at the freight terminal at Hams Hall, and this train is also bound for the Suffolk port of Felixstowe, with loco No. 66757 *West Somerset Railway* in charge.

Both Hinckley and Narborough are served by Cross Country Trains' hourly stopping services between Nuneaton and Leicester, with approximately every other service also calling at South Wigston. Three-car unit No. 170104 pauses on a Birmingham-bound service. Although these passenger trains are operated by Cross Country, the stations are managed by East Midlands Railway.

Knighton Junction, Swannington and Leicester Junction (KSL)

The next stage of our journey around the county sees us return to a point just north of South Wigston. A freight-only line between Knighton Junction, to the south of Leicester station, and Burton on Trent survives. Once a triangular junction, Knighton is now reached only from the south as seen in this photo of track machine No. DR74002, stabled on the curve, adjacent to the Midland Main Line. The northern chord of the triangle was removed in the 1980s.

This spur at Knighton Junction is often used as a convenient means of loading and unloading locos from road transport. On 11 May 2018, the curve is home to Class 27 locomotive No. 27059, formerly No. D5410.

A few months later, on 14 December 2018, it's the turn of No. 37207 to be stabled in the same sidings. This preserved loco had been in the hands of the Plym Valley Railway before being acquired by Colas Rail. This loco had become a celebrity in Cornwall, once carrying the name *William Cookworthy* in recognition of the man who discovered china clay in the county.

In this view of the sidings, the single-track branch to Burton on Trent can be seen on the left. This freight line to Burton runs north-westwards for about 30 miles, the majority of which falls within the county of Leicestershire.

The line was formerly known as the Knighton Junction, Swannington and Leicester Junction line (KSL). Its survival is mainly due to the aggregates traffic it carries from the important quarries in the area. On 22 July 2021, GB Railfreight's No. 66706 *Nene Valley* passes Bagworth on a rake of empty hoppers returning to Bardon Hill Quarry, near Coalville, following unloading at Colnbrook, near Slough.

Just to the north of Bagworth, a single-track spur serves Midland Quarry Products' plant at Cliffe Hill, also known as Stud Farm. Train loads of granite chippings leave the quarry by rail several times each weekday. This spur can be seen on the right of this view, looking north towards Coalville.

A mile further north, the KSL line reaches the substantial quarry at Bardon Hill. It is the headquarters of Aggregate Industries and is linked to the Leicester to Burton line via a short spur. On Saturday 7 April 2012, this is the view from the quarry spur line's crossing with the A511 main road. The stone workings were in the hands of Freightliner at the time, and a line of their Class 66 locomotives can be seen stabled for the weekend.

The company's privately owned shunters move the wagons around the quarry complex. On the morning of 22 July 2021 one of these is about to take a short rake of empties into the quarry for loading. Between 3 and 4 million tons of granite are produced annually.

On 22 July 2021, Aggregate Industries' 'Steelman' TH 297V is seen on the level crossing while running into the quarry.

Later the same day, GB Railfreight's No. 66765 has arrived at Bardon Hill Quarry with the returning empty wagons following unloading at Tinsley, in Sheffield. It is in the process of running round its train and is seen awaiting reversal at Bardon Hill signal box.

Following reversal, No. 66765 has now reversed on to its rake of wagons and is ready to haul them towards the quarry.

Most workings to and from Bardon Hill Quarry operate southbound in the Leicester direction. One regular exception is this working from Tinsley to Bardon. It is one of the few regular workings using the Bardon to Burton on Trent section of the KSL. It is seen passing Coalville's town crossing just prior to No. 66765's arrival back at Bardon Hill.

In recent years, few trains have regularly traversed the entire length of the KSL. Of late, however, rakes of box wagons have operated between Burton on Trent and both Bow, East London, and Acton Yard, in West London. In this view at Coalville, No. 66733 *Cambridge PSB* is working one such rake to Acton Yard.

The former Midland Railway signal box at Mantle Lane, Coalville, controls a section of the KSL as well as the holding sidings used for stabling rakes of wagons from Bardon when not in use. Passenger services on this Leicester to Burton on Trent line were withdrawn in 1964. Since then, there has been much talk of a reinstated passenger service between Leicester and Burton on Trent. There are, however, a number of potential obstacles to overcome. These include severe track speed restrictions along much of the line and lengthy block sections between the three manually operated signal boxes at Moira West, Bardon Hill and here at Mantle Lane.

Loughborough, including Brush Traction

With an estimated population of around 60,000 in the 2021 census, the town of Loughborough is the second largest in the county, behind the city of Leicester. It is at the northern edge of the county, close to its border with Nottinghamshire and 13 miles from Leicester. This view shows No. 156410 calling on a local East Midlands Trains' service.

A HST move with a difference on 13 April 2013: a circular move from Etches Park, Derby, involving two power cars, including No. 43467 and a short rake of coaching stock. They had recently been transferred from Grand Central to East Midlands Trains.

Another regular visitor to Loughborough is the HST set that forms the Network Rail Measurement Train. On 17 April 2017, power car Nos 43014 *The Railway Observer* and 43062 *John Armitt* head north on a working returning to its Derby base.

A variety of long-distant freight traffic uses this stretch of the Midland Main Line, including this longstanding working from Bletchley to Peak Forest. On 10 October 2014, the working was operated by EWS (now DB Cargo) with their loco, No. 60099, in charge of the empties returning to the Derbyshire quarry. This working is now in the hands of GB Railfreight.

Heading in the opposite direction on the same day, No. 66152 *Derek Holmes – Railway Operator* hauls a short rake of steel coils from Margam, South Wales, to Corby in Northamptonshire. This is another longstanding working that continues to be operated by DB Cargo.

In 2013, Freightliner's small fleet of Class 70 locomotives were often rostered to haul an Infrastructure service of empty box wagons from Crewe to Mountsorrel. On 1 May that year, the return loaded working from the Leicestershire quarry is seen just north of Loughborough station with No. 70015 in charge.

Another Freightliner working from Mountsorrel is seen passing through the station on 1 September 2014. On this occasion, No. 66524 is hauling a short rake of loaded ballast wagons to the yard at nearby Toton.

By the time this photo was taken, on 27 May 2016, the Crewe to Mountsorrel and return ballast workings were in the hands of Direct Rail Services. On this occasion, their Vossloh-built Class 68 locomotive, No. 68001 *Evolution*, was at the helm.

The town of Loughborough has been the home of Brush Traction, and its predecessors, for over 150 years. Their workshops are located close to the Midland Main Line, adjacent to the station. It has been a subsidiary of Wabtec since 2011 and provides maintenance services for a diverse range of railway locomotives and rolling stock. On 24 March 2015, for example, a Chiltern Railways driving van trainer, No. 82302, is about to enter the works complex. It has been hauled by West Coast Railway Company's English Electric Type 3 locomotive, No. 37706.

A GB Railfreight move on 1 September 2014 saw a pair of Class 73 locomotives, Nos 73005 and 73006, arrive at Brush's Workshops. The pair are about to take the single-track curve into the complex being hauled by No. 66728 *Institute of Railway Operators*.

The works carried out a major programme of upgrades to six Class 73 locomotives on behalf of GB Railfreight. These locomotives were renumbered 73966 to 73971 inclusive and fitted with the necessary equipment to work with the new coaching stock recently delivered for Caledonian Sleeper services in Scotland. This photo taken on 3 November 2015 shows No. 73006, now converted to No. 73967, within the works yard.

The Class 92 electric locomotives were built at Brush workshops in the 1990s. The two examples in this photo, Nos 92006 and 92020, were both built in 1994. They are seen in the Brush Traction yard on 28 January 2017. No. 92020 has seen little revenue earning use, having spent much of the last two decades in store at various UK locations.

Bottesford (Nottingham to Grantham Line)

The Leicestershire village of Bottesford lies in the extreme north-east corner of the county. It retains its railway station, on the Nottingham to Grantham line. On 14 July 2021, East Midlands Railways (EMR) No. 158777 passes the station on a Liverpool to Norwich service.

The same day, a pair of EMR's Class 156 units, Nos 156403 and 156405, call on a Skegness to Nottingham service. The station is geographically isolated from the rest of the county's remaining railway stations.

Barrow upon Soar, Sileby and Syston

Between Loughborough and Leicester, these three passenger stations, which closed in 1968, were reopened in 1994. The large village of Barrow upon Soar is 3 miles south of Loughborough. Its railway station was resited on its reopening, and this is the view from the northbound platform looking towards Leicester.

An hourly passenger service, operated by East Midlands Railway, links Leicester, Loughborough, Nottingham and Lincolnshire, serving almost all stations each hour, including Barrow upon Soar.

The three stations were reopened with 'Ivanhoe' branding. It was envisaged that this would form the first phase of a much grander scheme to reinstate an extended passenger service through Leicester serving new, reopened stations on the freight-only line towards Coalville and Burton on Trent. This has failed to materialise to date. Curiously, Willington station, between Burton on Trent and Derby, was also reopened in the mid-1990s, and retains Ivanhoe branding.

A variety of non-passenger traffic also uses this stretch of the Midland Main Line. On 29 July 2021, No. 66724 *Drax Power Station* is seen heading north through Barrow upon Soar on a light engine move from GB Railfreight's yard at Wellingborough to their depot at Doncaster.

Sileby is about 4 miles south of Barrow upon Soar. Like Barrow, it is unstaffed, its facilities are basic and the entrance and exit down to and from street level is particularly uninviting. On 12 April 2018, No. 156413 pulls away on a northbound service.

East Midlands Trains' four-car Meridian, No. 222104, passed through the station moments later on another northbound unidentified working.

Once a common sight on Britain's rail network, this rake of DB Cargo's HTA coal hoppers made a rare appearance in the area in April 2018. Their Class 66 loco, No. 66138, heads south through Sileby on a Scunthorpe to Margam working.

Heading in the opposite direction, GB Railfreight's No. 66782 is seen working light engine between the yards at Whitemoor and Toton.

The final stop between Loughborough and Leicester is Syston, 3 miles south of Sileby and approximately 4 miles north of Leicester. The station buildings have long gone, but, unlike Barrow upon Soar and Sileby, the station at Syston offers step-free access via a simple ramp at street level.

Syston has a single, bi-directional platform. On 29 July 2021, an East Midlands Railways (EMR) Class 158 unit, No. 158813, calls on a southbound service to Leicester. The single-track section, or 'slow' line, runs from here to just north of Leicester station.

DC Rail's Class 60, No. 60028, was following south waiting to enter this single-track section. It is returning from a light engine move from Leicester to Chesterfield and return. The line eastwards towards Peterborough curves to the right in this photo.

Freightliner's Class 66 loco, No. 66525, comes off the curve and approaches Syston station on a container train from Felixstowe. Having run across East Anglia to Peterborough, the train will join the West Coast Main Line at Nuneaton, then head north to Crewe.

The local passenger services have historically been predominantly in the hands of EMR's Class 156 and Class 158 units. More recently, their recently acquired Class 170 units have started to make appearances. Sporting its newly applied aubergine livery, No. 170532 calls at Syston on a Leicester-bound service.

Mountsorrel

The Leicestershire quarry at Mountsorrel is situated less than a mile south of Barrow upon Soar station. It is operated by Tarmac, one of the area's most important railway freight customers. On 21 July 2016, the company's shunter, named *Roy-Don-Clive*, stands alongside the wagon-loading equipment.

The company also uses Zephir railroad locotractors for quarry wagon movements. On the same day, their machine is seen during the loading of a rake of IOA box wagons.

These wagons had arrived behind GB Railfreight's No. 66720, which is seen waiting in Mountsorrel sidings during the loading process. It will later return the loaded rake to Whitemoor Yard in Cambridgeshire, for use by Tarmac's major customer, Network Rail. The quarry supplies an estimated 70 per cent of all Network Rail's ballast needs.

Tarmac, Mountsorrel, is an important customer for DB Cargo. The rail freight operator is responsible for the movement of millions of tons of stone from the quarry to a variety of destinations across the country. This is particularly true in southern England, where there are no such hard rock reserves. On 4 July 2016, No. 66019 is waiting to couple up to its wagons for a loaded working to Radlett, Hertfordshire.

Direct Rail Services also operate ballast workings on behalf of Network Rail. On the same day, their Class 68 loco, No. 68020 *Reliance*, waits in the Mountsorrel headshunt siding. It, too, will shortly leave with a loaded rake on behalf of Network Rail, this time to the yard at Basford Hall, Crewe.

Humberstone Road Sidings, Leicester

These sidings, to the north of Leicester station, are frequently used for the stabling of trains waiting to enter the Mountsorrel quarry yard itself. GB Railfreight's No. 66730 *Whitemoor* waits in Humberstone Road on a rake of empty box wagons from Eastleigh to Mountsorrel.

On 13 April 2018, these sidings were home to a more unusual occupant. The sidings were used to stable the DB Cargo Management Company Train. On this occasion, it was hauled by their Arriva turquoise-liveried Class 67, No. 67002.

Melton Mowbray

The line towards Stamford and Peterborough leaves the Midland Main Line at a triangular junction at Syston. Heading eastwards, it continues to serve passengers at both the Leicestershire town of Melton Mowbray and the Rutland town of Oakham. The station at Rearsby, closed to passengers in 1961, was one of a number of smaller stations closed during the mid-twentieth century. The station building survives in private use.

On 14 July 2021, three-car Cross Country unit No. 170105 passes the closed station while working a Birmingham New Street to Stansted Airport service. An hourly service in each direction is currently offered to passengers between Leicester and Peterborough.

This cross-country line is used by a number of freight trains each day, including intermodal container services to and from the Port of Felixstowe. DB Cargo's No. 66148 *Maritime Intermodal Seven* heads eastwards past Rearsby on one such working from the container terminal at East Midlands Gateway.

On the same day, DB Cargo's sister loco, No. 66124, approaches Rearsby heading in the opposite direction. Sporting the company's revised livery, it is working from Felixstowe to East Midlands Gateway. It will take the north curve at Syston on reaching the Midland Main Line.

The station at nearby Frisby also closed to passengers in 1961. On 14 July 2021, No. 66112 passes the signal box at Frisby while working a rake of empty steel wagons from Corby to Margam, in South Wales.

This view across the fields near Frisby sums up the rural aspects of this line. Colas Rail track machine No. DR98008 is seen heading west a couple of miles from the village of Frisby on the Wreake. It is on a Welwyn Garden City to Rugby working.

To the west of Melton Mowbray, a 13-mile stretch of the former Midland Railway route between Kettering and Nottingham survives. Leaving the Syston to Peterborough line at Melton Junction, this line runs for 13 miles northwards to Edwalton and is often referred to as 'The Old Dalby Test Track'. Its depot at Asfordby is seen here with Nos 56312, 313121 and Greater Anglia unit No. 720518 amongst its occupants.

This complex is now known as the Melton Rail Innovation & Development Centre (RIDC), Network Rail. It offers a safe environment in which to test a variety of new and modified rolling stock. On 14 July 2021, two new London Northwestern Railway units, Nos 730101 and 730103, were present for such testing purposes. In April 2021, GBRf signed a four-year contract with Network Rail for operating this site on its behalf.

In July 2021, two resident shunters were based at Asfordby. Nos 08580, built in 1959, and 08922, built in 1961, can both be seen in this view. Grand Central diesel unit No. 180108 was also present, having arrived from Crofton, West Yorkshire, earlier that week. In the distance another pair of unidentified Class 720 units for Greater Anglia can be seen.

Access to the RIDC from the west involves a reversal, usually in the platform at Melton Mowbray station. On 9 September 2019, Europhoenix-liveried No. 37608 *Andromeda* is seen in the Melton station platform on one such move. It has just arrived from the company's base at Leicester to collect an outgoing unit from the test track.

With a population of around 30,000, the town of Melton Mowbray enjoys an hourly passenger service both westwards, to Leicester and Birmingham, and eastwards to Peterborough, Cambridge and Stansted Airport. In September 2019, No. 170108 calls on a Stansted-bound service.

On 3 December 2013, a pair of DB Cargo Class 60s, Nos 60100 and 60066, head west through Melton Mowbray on a test run from their home depot at Toton. This circular test run operated via Peterborough.

On 6 September 2019, No. 66531 heads west through Melton Mowbray with a Freightliner service from Felixstowe to Crewe. It will be routed through Leicester and join the West Coast Main Line at Nuneaton.

The same day, No. 66747 *Made in Sheffield* also heads west on a GBRf-operated Infrastructure service. The locomotive sports a unique Newell & Wright livery and is taking a rake of box wagons from Whitemoor Yard to Mountsorrel Quarry for loading.

Oakham and Rutland

The county of Rutland is the smallest in England and has a long, close affinity with its larger neighbour, Leicestershire. The county once boasted ten passenger stations but only the county town station of Oakham has survived. On 14 July 2021, No. 60028 passes the village of Ashwell with a route learner from Peterborough to Leicester. The loco carries the 'Cappagh Group of Companies' branding.

GB Railfreight's No. 66722 *Sir Edward Watkin* passes the semaphore signals and signal box at Langham Junction, about a mile from Oakham. The loco is working a westbound intermodal from Felixstowe to Birch Coppice, West Midlands.

Railways of Leicestershire in the Twenty-first Century

To the east of the signal box, four tracks lead towards Oakham station, about a mile away. DB Cargo's No. 66040 waits for the road westwards with a rake of empty box wagons returning from Chesterton Junction, near Cambridge, to the quarry at Mountsorrel.

Oakham station is situated halfway between Leicester and Peterborough, approximately 25 miles from each. The station dating from 1847, together with the nearby footbridge and signal box, are all Grade II listed buildings.

The station is managed by Abellio East Midlands Limited, trading as East Midlands Railway, although almost all passenger services are operated by Cross Country Trains. Their unit No. 170110 approaches Oakham station on a Stansted Airport to Birmingham New Street service.

DC Rail's No. 60028 is seen again on 14 July 2021. This time the loco is passing Oakham's signal box and town level crossing on another run between Peterborough and Leicester.

On 3 October 2013, DB Cargo's No. 66067 awaits the signal to leave the loop adjacent to the station. It is working a rake of loaded stone wagons from the Leicestershire quarry at Mountsorrel to Lafarge's terminal at Kennett, near Newmarket.

GB Railfreight's No. 66781 is seen heading in the opposite direction. It is working an intermodal from the West Midlands terminal at Birch Coppice to the port of Felixstowe.

The final photo from Rutland is taken close to its county border with Northamptonshire. The Welland Viaduct carries the Oakham to Kettering railway line across the river valley of the same name. It is approximately three quarters of a mile in length with eighty-two arches, each with a 40-foot span. Completed in 1878, the structure has the small communities of Seaton in Rutland and Harringworth in Northamptonshire at either end of its span.